AFTER DEATH

AFTER DEATH

FRANÇOIS J. BONNET

Translated by

AMY IRELAND AND ROBIN MACKAY

URBANOMIC

Published in 2020 by
URBANOMIC MEDIA LTD,
THE OLD LEMONADE FACTORY,
WINDSOR QUARRY,
FALMOUTH TR11 3EX,
UNITED KINGDOM

Originally published in French as *Après la Mort. Essai sur l'envers du présent*

BRITISH LIBRARY CATALOGUING-IN-PUBLICATION DATA

A full catalogue record of this book is available
from the British Library

ISBN 978-1-913029-70-8

Distributed by The MIT Press, Cambridge, Massachusetts
and London, England

Type by Norm, Zurich
Printed and bound in the UK by
TJ International, Padstow

www.urbanomic.com

CONTENTS

Nor is any life ever satisfied to live in any present, for insofar as it is life it continues, and it continues into the future to the degree that it lacks life. If it were to possess itself completely here and now and be in want of nothing—if it awaited nothing in the future—it would not continue: it would cease to be life.

Carlo Michelstaedter

1
ANAESTHESIA, AMNESIA

THE DOUBLE

I am finite. This is perhaps the one thing I can still say with some certainty, but it may no longer mean what it once did. Death is gradually being eradicated from finitude. Saying 'I am going to die' or just 'I am mortal' is not so simple anymore, not quite so self-evident. Such statements are less definitive than they have ever been. It seems inevitable that advances in biotechnology will soon force us to radically reconsider the traditional notion of death. The word will lose its unequivocal meaning. Already, the cryogenically preserved bodies of affluent death-dodgers sow doubts about death's finality. If these icy corpses were one day to rise and walk again, what would death mean then? A breach is opening up and a flood of questions is already rushing in: Will we be dead when all of our organs, bones, tissues, and fluids have been replaced one by one, when our old bodies have been superseded by new composite anatomies, assembled and reassembled one part at a time? Having regenerated ourselves several times over, will we be entitled to say, 'I have died multiple deaths'? How many duplications and alterations of our DNA will it take before we consider our original selves to have given way to new selves—the same yet different, like the ship of Theseus? Will we still be ourselves when our brains, and all of our thoughts and memories with them, have been modelled, reproduced, and uploaded, rendering our bodies obsolete? This isn't the reality we live in. At least, not yet. But even if, for now, immortality remains an unattainable fantasy, life extension technology seems to be getting ever more plausible, so that soon it will no longer be possible to define life as the simple mirror image of death.

However, even if we set aside futuristic narratives that take immortality almost as a given, and simply accept our condition as human beings living in the early twenty-first century—undoubtedly doomed to perish—it doesn't seem to make any difference: death remains out of reach, elusive, implausible, indistinct. Moreover, it is impossible for us to find a grounding in our own mortality. Death—or more precisely, *our own death*—falls outside the scope of our experience and as such remains unknowable. We can only confirm our intuition that we are alive 'for ourselves' through the lived experience of our existence. And the one thing that allows us to establish this sense of ourselves *as living beings* is the fact of our boundedness. None of the many attempts to establish in generic terms what it *is* to be alive are ever wholly satisfactory. Reproduction, homeostasis, and self-preservation are commonly cited as invariant properties of living organisms, but these functions only make sense in terms of the *interiority* of an organism. The integrity of a living being immediately implies its circumscription in space. For no organism can sustain itself without a boundary that gives it a form and allows it to interface with the world outside of it.

This boundary, this spatial limit, is something I simultaneously incarnate and grasp through my body; it is effectively all that I am. But this bodily limit is also something I must consign to secrecy each time it reveals itself, plunging it over and over again into obscurity so that a dream of my infinite existence can be perpetuated and expanded. Nevertheless, at some point I must reconcile myself to the fact that the first thing I am conscious of, as a living being, is the fact that I have limits. The limit of the surface of a skin that brings me pleasure and pain; the limits of my growth, my flexibility, my reach; the limits of my strength, my abilities, and my energy; the limits of my senses.

It is these limits, these boundaries, that constitute me as an individual. Self-consciousness is perhaps nothing but an awareness of being finite, of being a boundary between an inside and an outside. If the boundary is compromised and I lose my blood or my grey matter, if my finite-being is entirely opened to the outside, vaporised and dispersed, then my whole experience of the world disappears. My 'I' disappears. Without this boundary to demarcate interior from exterior, no exchange would be possible, and there would be neither interaction nor porosity. Everything would be there already, frozen in an undifferentiated, infinite, and eternal space.

Conversely, the finite character of existence directly entails the possibility, and indeed the *necessity*, of becoming and relationality. What is irreducible in my being is its finitude. And it is this finitude that opens it up to becomings. To delimit the space and time of one's existence is immediately also to circulate, move, and orient oneself *externally and in time*. If I am not omnipresent, if I am not eternal, then I can move and become—or rather, I *cannot avoid* moving and becoming.

This finite-being, the irreducible core of my existence, superimposed upon and coincident with my body, moves within the exterior space that is the world. This entails an extremely complex interfacing in which sensations, affects, and concepts are aggregated, organised, arranged in hierarchies, exchanged, and distributed among gods, humans, animals, and plants, as well as countless inanimate *others*, from stones to technical objects to stars.

Lived experience is therefore primarily grounded in the articulation between finite, centred being, and the multiple, potentially eternal and infinite world that surrounds it and embraces it. Infinity and eternity are not to be understood here as mathematical, spatial, or temporal determinations, but are meant instead to evoke the world's exorbitance—its omnipresence, the limits of which remain unknown. They therefore designate indefinite quantities, and are opposed to bounded, territorial, and finite-being in so far as they are potentially limitless.

Nothing is possible without this coupling of constrained being and immeasurable expanse. It constitutes the axis around which everything resonates and unfolds. This may seem obvious—a basic observation we have all made, more or less formally, but always with this coupling serving as the primary matrix of our relationship to the world. It is not at all obvious, however, and cannot be taken for granted. To put it plainly, this coupling is toxic. It leaks. The interface is unstable.

There is a doubling, a divergence between two vital forces: the centripetal force of the progressive, unidirectional existence of finite-being (that which, even today, lasts only from birth to death), and a centrifugal force that drives the multidirectional currents of the social world—those frameworks of signs that inscribe the individual within a non-oriented time and space that reaches out beyond them.

This doubling arises from the decoupling of two becomings that can no longer hold together, rather than from the confrontation of two antagonistic forces. But there is a dynamic instability here: one force overpowers the other and sucks it in like a vacuum. Bounded, finite-being, overcome, finds itself drawn into the activity and exchanges of the open world. The aim of this text is to show the consequences of this doubling, to render visible its impact upon our lives, to define its toxicity, and to identify its causes. And to bring this process of doubling to light, we shall have to reveal its effects.

So, what happens when the coupling that presides over the existence of individuals becomes unstable, when finite-being is dispersed into the infinity of the world? Why does the infinite world hold such an irresistible attraction for finite-being? To begin with, we can put forward one generic notion that covers a broad set of behaviours which manifest this disequilibrium: that of *sacrifice*. Sacrifice should be understood here as the pull of the vacuum, that momentum through which finite-being forgets itself or denies its own boundedness, to be dissolved in the limitlessness of worldly space, to become, exclusively, a being-in-the-world—just one being among others, indifferent to itself.

This category of sacrifice does not refer, then, to the practice of ritual expenditure that transforms the excess and pure loss of the gift into a tribute to 'solar' divinities. Nor does it pertain to a pledge of allegiance to God, as in the trials of Jephthah and Abraham. Rather, it denotes a radical endogenous operation wherein what is risked is one's own self. Sacrifice here is self-sacrifice—self-destruction for the benefit of everyone else. At first glance, this may seem rather an exceptional occurrence. On the contrary—it is the rule.

For there are countless degrees of self-sacrifice, from self-denial to actually killing oneself, with all the nuances of renunciation that lie in between. To sacrifice oneself, *to render oneself sacred*, is to terminate one's finite-being for the sake of the effects this will engender in the infinite world. Becoming a martyr by blowing yourself up in the name of your god, becoming a soldier and offering up your life for your country, rushing into a fire to save the lives of strangers while risking your own. But also—and far more insidiously—squandering your time and working yourself to death for the sake of the continued expansion of the company that employs you, enslaving your future to that of an abstract structure that supersedes you. With every sacrifice made, the vacuum exerts its power over finite-being and contests its right to exist.

Now, the value and legitimacy of a sacrifice may be assessed and debated within the infinite world, but the annihilation of finite-being is something that can never be subject to any evaluation. In sacrificing myself, I produce effects in a world that I cannot ever really be certain will outlast me, and which in any case will lose all consistency for me, since I will no longer exist. It's a truly impossible bargain. To sacrifice oneself is to act nonsensically, in a way that is literally absurd for finite-being. Yet this is a transaction that has been the basis of numerous civilisations. It is the original act of faith.

Sacrifice, then, heralds the final stage in the disjunction between two becomings, and the potential elimination of one in the name of the other. The dissolution of finite-being is not quite the same thing as the destruction of the individual, however, since the individual can very well flourish in its absence, simply allowing itself to be swept along by the becomings of others. When the disjunction is so extreme that finite-being is effaced, it is its very *relationship* to the world that is altered. This decoupling, this tearing of the individual in two—into a finite-being that constitutes it and a projected-being cast into a world that defines it—doesn't come out of nowhere. It is not a product of gravity or arbitrary terrestrial forces. It is the result of social apparatuses, the fruit of convergent ideologies, and even though its effects can be traced all the way back to the emergence of the very first societies, having always been employed in religious apparatuses and in the fabrication of transcendence, its near-perfection and widespread propagation is a recent historical achievement. Let's look, then, at the mechanisms that today occasion and sustain this division.

NO MORE FEELING

Many attempts have been made to establish a symptomatology of postindustrial society, and especially of of its spectacular aspects. What seems to emerge out of these different approaches, although paradoxically it is rarely mentioned as such, is a phenomenon that encompasses a broad spectrum of behaviours related to the dominance of the immaterial (images, knowledge, information) in social organisation. This phenomenon is *anaesthesia*. And this anaesthesia in fact offers another way to detect the division described above—the advent of a split between finite-being and social or projected-being.

Situated finite-being affirms itself through a confrontation with a potentially infinite and eternal spacetime. Projected-being, on the other hand, defines itself through the interactions it enters into with everything else. Anaesthesia does not result from the deepening schism between finite-being and projected-being but rather contributes to it. It is not a product of the division so much as one of the various mechanisms that may lead to it. We must therefore examine the conditions that produce and intensify this anaesthesia, and its immediate consequences.

But first of all we must determine exactly what we mean by anaesthesia. In the sense we are using here, it is not a purely physiological phenomenon, although the inevitable psychological changes caused by the decoupling of finite- and projected-being do indeed have immediate corporeal effects. But neither is it simply symbolic or metaphorical. It describes both a deficit of sensations and a growing inability to feel.

As such, anaesthesia, given its effects upon interhuman networks, could be defined as an inverted autism. Where autism inhibits the individual's ability to comprehend communication systems (verbal or otherwise) as well as their capacity to participate in and be absorbed into interrelational networks, promoting instead a 'pathic' relationship to things (i.e., a relationship based on intuition and immediacy), anaesthesia implies a dedication to the putting into words, exchange, and networking of what is perceived, at the cost of a more direct sensory relationship to it, a relationship with the world less invested in social signalling—or one less structured by the signs that surround it.

The anaesthesia of postindustrial societies translates, therefore, into an apathy toward anything that cannot be the object of an exchange, symbolic or otherwise. The logical evolution of the transformation of use value into exchange value has moved to the next stage here: it no longer just concerns commodities, raw materials, and concrete objects, but also knowledge and information. Of course, knowledge has always been used as a tool of domination, offering a strategic advantage to whoever possesses it. And of course it has always been convertible into tangible wealth (from Judas's thirty denarii to the fortunes made through industrial espionage and the sale of personal information collected by social networks). But something new is unfolding in the 'digital age': the becoming-autonomous of the economic circuit of information, which has less and less need to 'descend' into the material sphere in order to see its value appreciate, since it is now assessed on the basis of the *influence* and *hold* it exerts, whether measured by audience figures, traffic, or the 'buzz' that it generates.

And so the individual themselves becomes an agent in the circulation of information, sometimes as a producer, but mostly as a consumer. Caught up in the exploitation of sensory experience (sensory, because information, before becoming an intelligible 'immaterial substance', is always vectorised by a sense-complex, i.e. a set of perceptions that are witnessed and then reported), the individual begins to suffer a growing indifference toward any sense experience they do not want to, do not know how to, or cannot integrate into the flow of information. Over the last fifty years in particular, an entire information economy has been established thanks to the advent of telecommunications technologies. And along with it there has developed a whole new relationship to information, which is in fact a relationship to the world, operating through diverse channels but always relying on the same mechanism and bearing the same hidden message.

For example, the dream of the advertising world, where everyone is forever young, doesn't just suggest an excessive reliance on the logic of marketing that exploits the seductive nature of youth in order to stimulate consumption. It also expresses a preoccupation common to postindustrial societies: the displacement of limits and, above all else, of the ultimate limit (at least for the time being), that of death.

Every society shapes itself around what should be said and must be shown as much as what cannot be said and must be concealed. The symbolic theme of death remains a serious adversary of the logic of the infinite world, since it refers precisely to the terminal condition of each one of us and brings our finite-being into view, thus underlining the inextricably separate nature of human beings.

For behind the rise of the information empire there lies a secret dream of community. It may only be a community of 'remote individuals', but it is a community all the same, one that seeks to amalgamate everybody onto the surface of the sayable, an undifferentiated plane of expression that belongs to the world beyond being.

Society seeks Eternity. It enlists us all in this quest and encourages us to believe in the limitlessness of our existence, our abilities, and our youth. The exaltation of young bodies, the purity that surrounds them, their pristine condition, and their construction as glorified bodies living in a hyper-vital world all indicate the same objective: the symbolic eternalisation of the human body, or, to put it another way, its abstraction. Of course, youth isn't reducible to a mere ideological construction. And it does indicate a state of vitality that *is* desirable. But it is precisely desirability that this appeal to youthfulness never fosters. The promotion of youth in no way constitutes an invitation to enjoy it. The imagery is detached, and even though it may be right there in front of you, it is always simultaneously postponed. Because if youth implies eternity, then there will always be time to enjoy it later—which is to say, never. Youth, as such, must not be consumed, it cannot enter the domain of fulfilled desires, for then the young bodies would become tarnished and corrupted, and susceptible all too soon to an imminent, yet still distant, death. Youth is invoked less for what it can concretely offer than for what it can imply: the expulsion, to the most distant horizon, of the end.

This means that the role played by youth in the work of warding off death is especially revealing, for it is part of a diversionary technique that is derived from anaesthetic operations. This diversion distances the individual from their terminal state of finite-being and transplants them into a space structured by the immutable and artificial signs of a phantasmatic youthfulness, an omnipresent figure adrift in the empire of images.

One might think that anaesthesia, if it occults the sensible, entails an abandonment of the world, folding being in upon itself and shutting off its interface with the outside. But in fact, exactly the opposite takes place: in subjecting our sensory experience to a filter that retains only what can be rendered into a tradable signal, we forego any direct, pathic relationship to the world and to ourselves, as the latter become masked by images of sensations. This is why the sensory imagery of youth is always so sanitised and evacuated of any real pleasure. It *signifies* desire, but never actualises it. In this sense, we can say that *anaesthesia is a regime of sensory abstraction*. It corresponds less to an absence of sensation than to its predetermination by an interpretive grid that contributes to the conditioning of sensory information by administering sensation, framing it, and ensuring its exchangeability.

As we are subjected ever more vehemently to structural pressures and to the forces of homogenisation, simply being able to feel becomes an increasingly complex affair. Everything experienced by the anaesthetised individual is manufactured into a sign, and must immediately be invested in a social relation, otherwise it will be neutralised and dissipate just as swiftly as it arrived. This type of anaesthesia, therefore, does not result from a diminution of the sensible. It cannot be understood as a kind of sensory deprivation. Rather, it arises out of an indifference *to* sensation. And this indifference is by no means the consequence of a deficit; on the contrary, it results from an *excess* of sensory solicitations.

FORGETTING

Indeed, anaesthesia is brought on not by a lack of sensations but by a sensory overload that ultimately renders all sensations identical to one another. Sensory commerce has had as its direct consequence the proliferation, large-scale distribution, and accumulation of sensations. This first began to emerge with the production of cultural commodities and the development of a mass art—for 'masses' that never existed except as a phantasmatic marketing ploy designed to carry out the very massification it was supposedly a product of—and has since led to the ignition of an undying flare of flickering sensory information, whose legitimacy resides solely in the *fact that it appears*, perfectly instantiating the maxim of the spectacle: 'What appears is good; what is good appears.'[1] In all places and at all times, images and sounds are projected, presented to our eyes and ears at an ever-accelerating pace—capturing us, captivating us—across all communication channels. Moreover, the 'speciation' of the sensible into different carrier media is now being reversed, as the predominance of traditional media (such as radio, television, and print media), along with access to cultural commodities (such as music and films), shifts toward concentration into a single, globalised, audiovisual vector: the internet. This hegemony is all the more striking given that the usage of the term 'internet' appears to be in decline. It is increasingly rare to say that we are 'going' on the internet, or that we are connecting to it, since we are always already there—perpetually connected and subject to the solicitations of an uninterrupted flux of sensations.

Anaesthesia is therefore related not so much to the intrinsic quality of what is given to be felt, as to an absence of any *resonance* between sensations. The atrophy of the sensible that is anaesthesia is effectively caused by the increasing difficulty we find in allowing sensations, or more precisely

---

1. G. Debord, *The Society of the Spectacle*, tr. K. Knabb (London: Rebel Press, 2004), 9–10 (§12).

sensory experiences, to unfold over time, as the attention spans of perceiving subjects contract and the time spent on any one sensation becomes ever more fleeting. In the relentless sensory flux of networked postindustrial societies, one sensation displaces another before it has had a chance to 'bloom'. We turn away from one sensation to make the most of the next, which is just as likely to be neglected as the one that came before it, and so on. In this sense, the pervasive anaesthesia of the contemporary world is also a form of *amnesia*—a return to an infantile state where every event effaces the previous one, preventing the emergence of any kind of 'epigenetic' perspective, i.e. eradicating the possibility of coherence.

Indifference and incoherence turn out to be two faces of the same operational complex, where anaesthesia is derived from amnesia, and apathy tends toward forgetting. The motor driving this amnesia becomes more conspicuous the more the sensory flux accelerates, a snowball effect that owes itself primarily to the progressive overcoming of the technical constraints (bit rate and storage capacity) limiting the exchange of dematerialised sensory information. This acceleration is therefore caused by the well-known effect of entrainment, in which increased availability and access modify behaviours and create new needs (a strategy known as 'technology push'). It is quite striking to see how the generation born at the beginning of the twenty-first century, having grown up in the era of ubiquitous and continuous access to sensory flows, is developing advanced skills for the parallel management of simultaneously-appearing layers of information. These capacities have been heavily solicited and fostered by the 'multitasking' approach of the computational technologies that are now central to every operating system. However, this new ability to switch between flows invariably brings with it an increasing inability to restrict oneself to a single flow and follow it through to its end. The many bifurcations of the sensory current generate a powerful undertow that proves increasingly difficult to resist as they accelerate and become ever denser. Irrespective of time or place, layer upon layer, the multiple streams of a sensory realm free of all physical constraint drag us along with them.

Lately, a new threshold has been crossed by the film industry. Disney has been working on a concept they call 'Second Screen', designed for use in traditional movie theatres with the intention of generating 'added value' by augmenting the basic experience of watching a projection, which is apparently no longer an adequate substitute for watching a movie at home. The 'Second Screen' format requires the resequencing of existing feature films so that 'breaks' can be integrated into the narrative, during which interactive games related to the themes of the film can be played using touchscreens. The fact that this project is being undertaken by the great giant of the children's entertainment industry is quite revelatory: in an attempt to adapt to our youngest generation's habits of sensory reception and consumption, Disney is actively participating in a radical paradigm shift in access to the flow of information, which is to say access to the sensible itself; a shift that is inducing new behaviours among a young generation of consumers who, faced with the mediatisation of sensation, prefer action and interaction over dreaming and imagination. No doubt Disney fears it will 'lose' its audience if it doesn't continually grab their attention at the pace they are accustomed to outside of the cinema, outside the temporary parenthesis that watching a film has now become.

After all, access to the sensible is inextricably linked to the time allocated for its unfolding. The administration of time, or more precisely the *rhythmics* involved in the release of the sensations it fabricates, has become central to the culture industry, which has therefore started to devise temporal strategies to optimise the dissemination of its products. These strategies revolve primarily around two axes that seem contradictory at first glance but are in fact complementary: forgetting and repeating.

Forgetting is solicited more and more often, and is now being operationalised to create novelty. The tactics of starting over or rapidly reconfiguring a familiar situation are becoming more highly prized because they are increasingly feasible. The logic of the reset is now employed everywhere, but the film and video game industries are in the vanguard. In endlessly rewriting the same film with the same plot, the same characters (but not the same actors, although they usually embody the same stereotypes), Hollywood seems to have discovered a new resource to exploit: forgetting. A perfect illustration of this would be the astonishing number and the increasingly rapid arrival of reboots, especially of blockbusters, and their continuing success at the box office. At a lower cost and with less risk, Hollywood launches new productions based on old ones that have proven their value but that the public has already forgotten. A new era seems to be upon us: the era of the stuttering of the sensible.

Repetition surfaces as a complementary strategy to forgetting, for forgetting allows repetition free rein. As well as being a well-known method of captivation (from Pavlovian conditioning to advertising slogans to playing pop songs on a loop), repetition becomes a tool of temporal capture, as it 're-instantiates' perception multiple times over in the same way, and petitions it in a unilateral manner. Where repetition can operate as a matrix of difference so long as there is some degree of thickness or duration between two identical occurrences, making the moment of its reappearance into a moment of difference, amnesiac repetition, in so far as it eliminates any thickness between subsequent instances of its recurrence, erases all possibility of difference. For the amnesiac, there is nothing that is not caught up in the cycle of repetitions.

The barely concealed dream of these combined strategies is to reduce the sensory horizon to a single, monopolistic stimulus that can be repeated as many times as necessary before a 'new' avatar with the same qualities arrives—as promptly as possible—to succeed it. In other words, the technique of planned obsolescence that has been employed on a grand scale since the beginning of the twentieth century, from the automobile industry to nylon stockings to batteries for electronic equipment, is now also being applied to cultural products (i.e. to the sensible), churning through tighter and tighter cycles of appearance and disappearance until stupefaction sets in.

But, again, it would be a mistake to see these practices and tendencies—which are leading to a widespread homogenisation of the sensations arising from these modes of apprehension—as a primary cause. The technologically-enabled overburdening of the human sensorium that produces sensory amnesia is itself only an effect. The cause lies elsewhere, and it is only indirectly connected to the sensible relationship to things. Forgetting and repetition may have become sensory functions, tools for the erasure and rewriting of sensations, but they fulfil another, more profound function within the apparatus of the apprehension of time.

Repeating things and forgetting them has a dual purpose. Of course, repetition and forgetting ventilate the sensorial field, providing free spaces and a rhythmics appropriate to the consumption and distribution of cultural products. But at the same time—and this is undoubtedly what makes the whole operation work—by transfixing us, by anaesthetising us, they reaffirm an eternally identical, unalterable present, thus contributing to the decoupling of finite-being and projected-being. They therefore participate in the projection of being into a symbolic space that distances it from death.

As we have said, death is in decline. Advances in bodily reconstruction techniques and the modelling of human brain function are made every day, despite their risky prospects. Some are already preempting the moment when the body will be fully reconstructible, redrawing the boundaries between the living and the dead. What is properly backed up will live, and what is forgotten will die. If resurrection were actually possible, would a cryogenically preserved human being still be 'alive' if it turned out that, once returned from the dead, they had lost their memory? Or would that person have been replaced by a new, as yet unknown being, a biological blank slate? Should such things come to pass, forgetting will subsist even in death, taking over death's function—that of redrafting the limits of finite-being, of destroying its continuity and the permanence of its integrity. This operational equivalence between death and forgetting leads us to the following paradoxical state of affairs: by establishing forgetting as the primordial function for the warding off of death, we betray a deeper intuition of their equivalence, and effectively realise this equivalence. The forgetting of death becomes death by forgetting.

What becomes ever clearer here is that amnesia is not an end in itself. Beyond simply making us forget, amnesia makes us more *present*. Just as paradoxically, it appears that this enhanced presence corresponds to a deficit in the intensity of experience, and so, in a way, it corresponds to an *absence*. In a currency that is ever more current, everything has the same value as everything else. The postindustrial era's spectacular anaesthesia/amnesia is intimately linked to a domain of the present, the current, that is ever more condensed, ever more instantaneous. Amnesia has perhaps brought to light a new-found inability to disconnect ourselves from what is right in front of us, an inability to perceive in any way other than through a relationship with a hyper-current world.

2
THE PERPETUAL PRESENT: A LIFE SENTENCE

TAUTOLOGY

Everyone knows the Parmenidean tautology according to which being *is* and non-being *is not*. It is an irrefutable assertion. In establishing the law of identity, it lays the foundations of logic. It also harbours depths often masked by the dazzling self-evidence of its superficial meaning. And yet our lived experience contradicts this doctrine at every moment, as we are continually projected toward and traversed by absences, elsewheres, and inexistences.

There has always been a tradition of thinking what *is right now*, of reducing the real to the present, which has continued to develop over time, cutting its ties with pre-rational, still superstitious ideas that saw the nonpresent as an active power almost equal to the present. The Hedonistic and Epicurean traditions, in which pleasure is the pivot-point of existence, designated the present moment as the ultimate site of joy.

Epicurus's great coup was to have disarmed the infinite by declaring that the gods did not concern themselves with us, and that the only reality of our lives is an earthly one. Living in the present, therefore, became the sole purpose of existence. There is no longer a Hades, an Elysium, or a Tartarus. No heaven or hell awaits us, ready to take stock of our past lives and consign us to an eternity of happiness or suffering. Life on Earth is no longer the first stage of a test, the trial of existence, the only goal of which is to attain salvation in the afterlife. Nor is it the seat of sin and penitence—what Nietzsche called the 'corruption of the soul'.[2] By rejecting the reality of an afterlife, Epicurus resituates the totality of existence in the present, the only place where joy

---

2. F. Nietzsche, *The Anti-Christ and Other Writings*, tr. J. Norman (Cambridge: Cambridge University Press, 2005), 61 (§58).

can truly be attained. This notion of pleasures promised to those who know how to live in the present is expressed in a now famous formula: *Carpe diem (quam minimum credula postero)*, 'Make the most of today and don't worry about tomorrow.'

Horace's *Carpe Diem*, when identified with a pleasure-oriented, hedonistic mode of thought, becomes an even more misguided slogan. Originally meant to illustrate the Epicurean principle of relishing the moment in order to counteract the uncertainty of the future and the inevitability of death—which is ultimately a way of reconciling oneself with the precarious nature of existence—the usage of *Carpe Diem* that coincides with the rise of consumer society makes of it a sort of magical formula, erecting recklessness and inconsistency into the twin pillars of the present's sudden ascendency, affirming the present as the sole locus of power and desire.

In this way, the Epicurean doctrine is perverted and turned against itself. Immortality, Epicurus's rejection of which directly opened the way to enjoyment of the present, is resurrected by a logic of overinvestment in the *current* that endorses a denial of all becoming, of any inexorable end, with the capacity to project oneself elsewhere being perpetually scrambled by the presentation of the here and now. Where Epicurus makes the inevitability and finality of death the principal argument for living and taking pleasure in the present, postindustrial ideology uses the promise of immediate pleasure as a diversionary technique that tries to ward off death, its inevitability and its finality, by eternalising the present.

The invitation to make the most of the present is therefore no longer a philosophical commitment, a way of conducting one's life in the limited time one is given. It has become an endlessly iterated injunction serving the twofold objective of banishing death and making us better consumers. The cult of the present instant has established itself not only in the ideological field, but also in the operational flow of the presentation of events. To put it another way, the present instant has gone from being vectorial—a vehicle of ataraxic thought moving toward happiness and away from suffering—to being an autonomous generator of events, cut off from any ulterior movement and designed only to intensify the present.

The increasingly frenetic accumulation of sounds and images is a part of this intensification, aiming to fill the void of a present that is forever sinking into the past. At every moment, the world must be supplied with novel sensory material generated to replace existing material that has already become obsolete. The exaltation of the present as the promise of pleasure is therefore never really meant to be actualised, since it will immediately be obsolesced by a continuously renewed influx of promises. Needs, cravings, and desires are endlessly rebooted.

The present moment, as constructed by our society, is nothing but a procedure of forgetting, a catalyst for amnesia, and as such it enables the repetition which, in turn, through the presentation of novelty, allows us to distance ourselves a little further from the past. And this writing and rewriting of the present—as if it were a palimpsest—continues to accelerate, demanding that we forget at an ever higher frequency. What *was* is of no concern; all *has beens* are eliminated in favour of what *is*, with little regard for what *will be*. Where the modern era anticipated a present yet to come, transposing current momentum into future accomplishments, our postindustrial, postmodern era sees the future as nothing but a vague, indeterminate site hosting an indistinct cloud of promises and signs as desirable as they are deadly.

Our spontaneous attitude toward the future is now one of caution. The enthusiasm of the moderns has gone.

Individualisation has eroded the promise of the future. For the future involves the hopes of the community, but also despair at the inevitable end of each of its individuals. So community is no longer forged around a common destiny. The light of all the promises that shine out before humanity fades in the eyes of the one who already knows that they will not be there to see these promises fulfilled. There is no brighter tomorrow for the future dead.

And so we arrive at the stage of 'tautological living'—the instant consumed for its own sake and for itself—plunging us into perpetual forgetting, saturated as we are with presentness. This bombardment profoundly changes our understanding of the cycle of the presentation of events. Repetition, forgetting, and immediate disinterest in what has just been consumed are now constitutive of our life experience.

This pervasive amnesia—this stupor—is not then simply something that is passively undergone. It becomes integrated into mechanisms of power that rapidly learn how to take advantage of it, endlessly starting over, reconfiguring the political playing field with ever increasing frequency. The logic of the reboot and the reset is constantly brought into play. Ruined reputations are swiftly rebuilt—after a little time out, a public figure returns renewed, rehabilitated, back in favour. And the media also contributes to an increasingly tenacious amnesia, or more precisely helps make it easier to accept. The injunction to forget not exactly the past but the nonpresent, has proved most profitable. It is necessary to forget in order to fluidify the present, in order to be able to accommodate new yet forever identical discourses, to be able to reinstate some celebrity or other who has made yet another comeback—in short, to make the presentation of the same, of repetition, into an original, immaculate moment.

Thought itself has become journalistic. The 'thinkers' of today belong to the ranks of journalists. It is they who are now supposed to secure meaning in an undifferentiated flow of information. It is they whose task it is to think a real that has now dissolved into the current. For what makes the present 'present', confirming it in its tautology, is information, the apparatus through which events are presented. The consecration of the information age, which corresponds to an acceleration and an expansion of modes of access to information, attests to a tightening in the weave of time. The less friction encountered in the deployment of events, in their being brought to everyone's attention, the denser the present becomes. Events are immediately effective, rapidly shared, and globalised. They generate instant reactions which have their own repercussions in turn, and so on *ad nauseam*. Decisions, reactions, the taking of positions, counterattacks, and commentaries in the wake of any given event can no longer be postponed or delayed. Waiting is increasingly inconceivable, and more and more unbearable when it does happen.

The injunction of the present manifests itself as a summons to be always on the alert, permanently available, and ready to react. The injunction of the present is therefore an entreaty to interact with it immediately and to throw oneself into an endlessly narrowing present. This relation to instantaneity demands an ever more fine-grained subdivision of *present instants*, instants in which one is *present*, and which constitute the irreducible grain of temporal immediacy.

This is because 'tautological living' acts as a narcotic: by bombarding us with 'nowness' it anaesthetises any anguish we feel about the future that harbours our inevitable demise.

The injunction of the present, transmitted in every one of its twisted, tautological messages and slogans, has no other role than to postpone the death of each and every one of us to the farthest point—indeterminacy or inexistence. We must stop time, pluck from it ever more atomised clusters of instants, cast them up in the air, then weave them through the weft of eternity which, like a meagre blindfold, hides from our eyes the inexorable leaking away of the time in which our finite-being is embedded. Eternity is the present without shadows. An instantaneous, tautological, endlessly renewed present. A static time in which nothing can happen.

SYNCHRONISATION

In Christopher Priest's novel *Inverted World*, a strange city named 'Earth' slowly rolls along on rails, indefatigably pursuing a moving point called the 'Optimum'. On Earth, time is measured by distance ('I had reached the age of six hundred and fifty miles', declares the narrator at the beginning of the story). The Optimum leaves the past behind it, condemned to undergo strange spatial distortions. In front is the future, from which one returns aged. Over and over again, initiates of the Earth guilds lay down and tear up the tracks upon which the city advances, drawn along by cables and pulleys that are perpetually being installed and dismantled in pursuit of a perfect and unattainable point.

The Optimum is a figure of the hyperpresent. A paradoxical point that continually splits into two: at once the most tangible point, the radical here-and-now, and the most fleeting, irreal, and ungraspable point, because it is always already elsewhere. This point is not at all imaginary or symbolic. If, unlike in Priest's novel, it stayed completely stationary, it would nonetheless still be a figure of the incremental progression of every thing from one state to the next. The Optimum is that point, that present reduced to the instant, that projected-being ceaselessly pursues. For, just like the city-dwellers of Priest's 'Earth', what postindustrial society is actually seeking through this obsession with the present is to be permanently *synchronised* with the present, to be continually and always in phase with the optimal point.

In this way, the perpetual fascination exerted by the instant becomes the dominant way of experiencing the real. Dominant because its demands become more closely packed the more the 'grain' of the present (that is to say, the optimal point) is reduced to a pure instant. The hyperpresent qua present without duration, forever renewed, contains within itself the necessity of its own constant re-presentation.

Having become characteristic of the culture industry of the last fifteen years, the reboot is once more the obvious example of the dominance of the twofold procedure of forgetting/repetition. For it is no longer a matter of following through, of continuing a story or taking it up as it is. It is a matter of rewriting it so that it is both different and identical at the same time. What changes is the representation of the story itself: it has to take on the formal appearance of what is most 'up to date', a sort of composite veil allying technical performance, aesthetic tendencies, and dominant ideologies. And this formal up-to-dateness has no ambition other than to be transparent. For it is less a question of flaunting an aesthetic modernity than of avoiding being seen as dated, or as belonging to a specific epoch. The 'cosmetic' strategy of the reboot aims to endow whatever it reboots with atemporality—that is to say, an absence of any temporal inscription.

So the reboot effaces the past, but since it forever rewrites another story which, in the end, is always the same one, it is incapable of inducing any kind of anticipation. It rejects both nostalgia and futurism so as to draw as close as possible to the radical instant, the foundational moment of the hyperpresent. What is pursued in the logic of the reboot is nothing other than synchronisation. And all synchronisation must continually be verified against points of synchronisation. It must be calibrated continually.

The reboot, as a technique of resetting—resetting *to zero*—participates in the synchronic establishment of the sensible world, joining the struggle against the duration of things, aiding in its exorcism or denial. The experience of time then becomes less and less continuous, more and more discretised into instantaneous events which are no longer even micro-epochs. Time is sequenced in present instants that are no longer successive but substitutable one for the other—no longer a horizontal series but a vertical stack. In a world increasingly dominated by reboot logic—which moreover is not limited to cultural goods, but extends to the (brand) images of everything that can possibly produce value—the instant becomes a moment of verification and resynchronisation. Repeat, forget, restart. Such are the three phases of the regime of the hyperpresent, a whirlwind of instants in constant synchronisation which redesigns our relation to the real by accelerating it.

The passer-by—walking the streets with an uncertain step, absorbed by their smartphone, eyes glued to the screen, field of vision reduced to this single luminous surface, constantly at risk of bumping into a lamp post or another person equally captivated by their own screen—isn't living in some sort of simulated, virtual space, or lost in a dream world. It is true that their immediate surroundings no longer mean anything to them, any more than the cosmos, or the centuries of which their life is only a minuscule part, and which manifest themselves in everything—trees, swatches of sky, architecture, perfumes.... It is true that they don't really know where they are unless their device tells them, that they will walk along without noticing anyone and without anyone noticing them, short of an actual collision. And yet, for all this, such an individual is not 'off in their own world' or 'out of the world'. They are, on the contrary, precisely 'in the world'—that is to say, as close as possible to the real, actualising it perpetually by updating it via their synchronisation with the Network. What makes things 'real' for them is less their immediate surroundings than the uninterrupted exchange of data and information they send and receive as they interface with humanity as a whole.

In synchronised life, projected-being gains the upper hand over finite-being. The experience of life is now actualised and realised more by flows of signs than by sensory input from the surrounding world. Participation in the world is established through the social bond, in its most radical sense. The relation to things tips gradually into insignificance, ushering in a widespread anaesthesia. The spectacle of a sunset, for example, will only assume its full value once it has been captured by a photograph addressed to and received by a community.

Rituals of synchronisation have always existed. Whether in festivals marking the cycle of the seasons or solstices such as Saturnalia or Midsummer, birthday celebrations, or rites of passage, one's life is measured out by events, coordinates that serve to inscribe the individual in a time that goes beyond them and of which there is a common experience, thus constructing a proto-synchronous community. Indeed, it is this placing in common of the consciousness of the past and the consciousness of inhabiting the same present that founds the very possibility of the sharing of sensory experiences.

But the radical change that has taken place since the advent of information technology and the era of the network society is that synchronisation has become *permanent* and has lost its rhythmic function. Rather than marking and designating specific moments whose convocation (or *calling*, to follow the etymology of 'calendar') allows for the establishing of contemporaneity between people, synchronisation has multiplied such moments of verification to the point of bringing about an uninterrupted synchronous flux.

Being synchronised means staying informed or being 'in the loop'. It is a way of affirming one's connection to the global community of human beings. It means being a member of the Human Network. But although the predominance of the current may seem to realise the unconscious dream of a synchronous community, the flows of communication are now not so much designed to share information as to make sure that each individual feels constantly synchronised. In other words, synchronisation has become the ultimate form of communication: a *communication without object*. The only object that remains, if there still is one, is the present itself, a present that is verified at every instant. Only one community is possible, then: the community of the confirmers of the present.

The synchronous community inscribes the individual—finite-being—within a wider destiny, projecting them into the atemporality and limitlessness of a hyperpresent that is, precisely, the confirmation of the present. Now, it is in confirming the present that we prove our own *presence*. We exist, we live in a shared time, a time that has no duration and is always in the present tense. Being synchronised with the rest of the world thus comes down to huddling tightly together, amid the vast blackness of time, at that blind and eternal point that is the present.

The synchronised hyperpresent solves everything. It contests the finite nature of the individual by projecting them into a present that is perpetually screening and that demands their attention at every instant. It then expatriates them from that finitude, hooking them up to a synchronous community that reaches beyond them and extends them indefinitely.

Synchronisation is thus the deeper reason behind decoupling, and what is at stake in it. Finite-being cannot be totally synchronised, because it is inscribed within and consigned to its own temporality, a biological temporality, so to speak—whereas projected-being is now wholly aimed at verifying, at every moment, the present instant. Decoupling happens when projected-being enslaves finite-being by trying to synchronise it (that is to say, by refusing it its own rhythm). The paradigmatic decoupled being might then take the form of the young Taiwanese man who died after forty straight hours of online gaming at an internet café. Immersed in the eternal present, he was pushed beyond his physical limits, stifling the finite-being that constituted him, reduced to nothing but a projected-being in the synchronous world.

We must, at this stage, trace the modern history of the synchronous community and how it developed in parallel with technical innovations. And we should begin with radio. Radio allowed everyone to have the experience of listening to an event in its nascent state—that is to say, an event as it is happening *now*, even though it is happening *elsewhere*. With radio, synchronous experience is no longer an experience linked to simple co-presence. It becomes autonomous, and mysterious. This history, after an intermediate passage through systems of *instantaneous capture*, in particular cinematography, which allowed for the potential intensification of every instant as a decisive moment forever fixed on film, would logically have to continue with the birth of live television, which makes the image an even more powerful vector for the experience of instantaneity. For it is live TV that ignites our fascination for flows of synchronous images. The desire for immediacy, then, is attached to the scopic drive: images *represent* the present instant and are the guarantor of it.

This short history of the synchronous world would conclude, for now (until the hypothetical reign of 'telepaths'—'augmented' humans who would constantly be connected to one another without the need for an external interface), with the determinative influence of the internet on the ways of life of contemporary society, especially since its integration into mobile phone networks, which served to open up a state of almost permanent accessibility. In the space of a single generation, modes of synchronisation have been revolutionised, going from a unidirectional synchronisation (from transmitter to receiver) to a multipolar system where every connection also operates as a verification point for the present instant. The pace is no longer just accelerated by the competition between channels constantly clamouring to be as close as possible to the instant in order to capture a maximum of attention (as is still the case with the most recent avatar of live television, 'rolling' news channels), rather, it is attention to events itself that defines the obsolescence of the present and the need to constantly update it.

This constant need for 'updates' of the present is not only manifested through the perpetual refreshing of flows of data, it also invades the landscape of the sensible. Even if such a distinction is less and less meaningful (data itself can now materialise as a sensible object, from a simple alarm sound reminding us of a meeting to the complex procedures that allow us to create visualisations of the earth's surface), there is indeed a complementary rhythm that endures in the network era without being completely assimilated by it: the synchronisation of the sensible through cultural commodities.

Whether in fashion, validating and invalidating the sensory current through sartorial trends, or in music, with the latest hits replacing one another identically, the goal is not so much to express a vision of the world or to defend an aesthetic as it is to affirm, through fashion or music, one's belonging to the current, to the present. Attaching yourself to the present moment is a way of assuring yourself that you are not outdated, abandoned, left behind in the wake of the Optimum, doomed to disqualification and obsolescence. The culture industry, and indeed the fashion sector, are in fact nothing other than precursors of synchronous globalisation, or more exactly of the sensory synchronisation of the contemporary world.

Now, this synchronisation of the sensible is precisely an anaesthetic mechanism in the sense that what is at stake in sensory projection (the manifestation of a cultural commodity) is no longer sensory experience itself but an experience of synchronicity. It matters little what one is listening to, what one watches, so long as it is current, new. Furthermore, the concept of novelty no longer possesses any prospective qualities, and hasn't for some time now. The new is not the avant-garde. It is the present in its purest form. Novelty is a present that is not yet eroded by duration.

The condition of being current is thus the dual condition of amnesia and anaesthesia. Nothing lasts here. All that counts is the Optimum, the synchronous regulation of sensory experience. Anaesthesia and amnesia rely on a collapsing of the present, on a tautological operation, on stuttering, on a forgetting of sequences, and on a permanent striation of events.

The present becomes a lytic mechanism, pulverising the world into isolated objects that can no longer be brought back together into a shared becoming.

PERIL

The networked anaesthesia of the postindustrial era is tied initially to the influence of the present, locked in the grip of an ever more condensed, ever more instantaneous experience of the current. This kind of condensation, the acceleration of the present through the runaway presentation of synchronous instants, contributes to the onset of a generalised paralysis of the current. Everywhere, it becomes viscous. It coagulates. It accelerates so much that it begins to seem static. And as a generalised stasis emerges, so also does an indifference to whatever may happen.

Because, to tell the truth, nothing happens anymore. Nothing any longer has the time to happen. There is no duration left for anything to unfold in. Nothing can anchor itself in the world long enough to make sense. While the present still has a duration, the hyperpresent no longer does. It is pure currency, pure instantaneity. In the age of the hyperpresent, nothing happens that is not already undone by the event that comes immediately after it. In this inexorable stream of aborted moments—of instants separated from one another, shattered into ever more tenuous fragments, we find ourselves submerged. But this submersion is not just suffered passively. It is also provoked, sustained, and even desired by an implicit adhesion to the amnesiac project of the hyperpresent. Self-forgetting occurs through surrender—willing abandonment to the immersive force of the flow of synchronisation. We happily allow ourselves to be taken over by a microfragmented temporality that paralyses us, that inscribes us and submerges us in an eternal present. And if it is a general feature of postindustrial societies, societies of triumphant information, that we 'no longer have time', it is not really because things are going faster and faster, but because time itself has been dissolved into a multitude of futureless instants deprived of becoming, autonomous and independent of one another, and connected by a single, tenuous thread—the thread of radical currency, of the static hyperpresent.

This acceleration of the fragmentation of time is not the same thing as an acceleration of time. On Earth, there are still twenty-four hours in a day. But these twenty-four hours are broken up into a profusion of instants, all of which demand our complete attention before disappearing into the limbo of forgetting. For the pulverisation of time, and the stuttering that results from it, generate a simulacra that wards off and consigns to desuetude the maddening flight of time that leads us inexorably to our limits—to our end.

The maintenance of such stasis, however, is far from painless. We have entered a new age of sacrifice—from the innumerable cases of those who 'burn out', unable to bear the ever-increasing pace of their fragmented lives, to the casualties of the internet café who offer themselves up to the Network, having forgotten their finite-being in favour of their projected-being. Each of us sacrifices some portion of our lifetimes qua finite-being to the communal time of projected-being—a frozen, instantaneous time in which becoming is abolished.

The stasis of the hyperpresent installs a time that both exceeds us and diminishes us—the floating time of a world without future and without past. Time's arrow, so to speak, has been banished. If time has a direction, this is the case only in so far as it can be apprehended by the living, for its direction is actualised solely by finitude. But the hyperpresent, as much as it can, catches and suspends the unfolding of time in order to make of it a dead time. The toxicity of the hyper-present lies precisely in this suspension of becoming—an extinguishing of temporal resonances that leads to the negation of the living being itself.

For there is indeed an exhaustion of vitality that accompanies the impossibility of keeping up the pace, a burnout that comes with continually having to synchronise ourselves—to the latest news, the latest request, the latest order, the latest counter-order, the latest fashion, the latest music, the latest film, the latest star. As we have said, these things have no importance in themselves. They are just codes, passwords that allow us to follow the thread, and to get to the next synchronisation. One step missed, one slipped link, and we find ourselves lagging behind, obliged to redouble our efforts to catch up with the Optimum and resynchronise ourselves. Meanwhile, the sense, the substance, the purpose of the things that serve as synchronisation points—these have become almost entirely irrelevant, so much so that they express nothing other than the fact of their synchronicity.

In order for these synchronisation points to fulfil their synchronising function, they need to be easily identifiable. So even if the flows of information, images, and sounds do not *in themselves* signify anything, they must nevertheless *signal*. They perform the role of a catalyst, mobilising what is already present in projected-being. Little by little, audiovisual flows are reduced to simple activators of pre-memorised meaning, and as a result almost nothing truly original, unprecedented, or innovative can be expressed in the spectacle of instantaneity. Everything must already be given.

A somewhat imperial note is struck by this quest for immediacy and instantaneity, and the need for discrete elements to coordinate and compress time in order to satisfy the demands of synchronisation; an imperialism bent upon manipulating the sensible and reducing it to the mere signalling of the present. The reboot, as discussed above, is nothing other than this: starting over, saying the same thing as before, optimising the bringing into phase of projected-being with synchronous and global stasis.

As a result, we discover an increasingly tenacious inability to detach ourselves from these instant-objects so perfectly fashioned to attract our attention. The possibility of seeing in any other way than through this vision of the hypercurrent, tautological world grows ever more slight. Repetition and simplification thus contribute to a triumph of the present achieved through conservatism and the optimisation of the sensible.

This optimisation takes the form of a simplification of signifying exchanges across sensory vectors. The standardisation of cultural production has no other mission. For it is not so much a matter of using standardisation to homogeneously establish a dominant ideology, as one of reducing meaning to its bare minimum, to a mere function of recognition and synchronisation.

The simplification of exchanges encourages their proliferation, producing in turn a further tightening of the synchronous mesh. The rise of multitasking logic is a new avatar of this imperial strategy of the reduction of processes for the consumption of the sensible. The ability to interface with many things at once, switching between them almost simultaneously, has developed over the last fifteen years through the expansion of the discipline of informatics. Informatics, a nearly obsolete term whose original meaning (the automated treatment of information) we have forgotten, has become, moreover, less a technique specific to electronic calculating machines than a ubiquitous procedure for the interfacing of individuals with the real. The impact of this apparatus has evolved rapidly, to the point of its acquiring an almost hegemonic status. The first generation of 'digital natives' has fully internalised the fractalisation of temporality, and in fact has an experience of time that is already different from that of the preceding generation, which had experienced the inertia of communication systems in which most of the time one was unreachable—that is, desynchronised.

Have we arrived at the epoch of temporal capitalism, where instants are refined, multiplied, and accumulated to be exchanged in simplified, synchronous transactions with the sole aim of maintaining the paralysis of the hyperpresent and the structural forgetting that it generates? The grip of the present flourishes into an incessant synchronous updating designed to stick as closely as possible to a dreamt-of real that is stable because it is continually reinstantiated. It is a dictatorship that binds experience to its rhythm, forcing it to express itself in the instant, refusing it the right to unfold in duration. It thus robs experience of all becoming and complexity.

For the experience of the instant cannot be traversed (that is to say, lived) unless it is redirected toward a becoming. How can we feel when every instant is self-sufficient, when nothing comes or goes, when all perspectives and horizons have been eliminated? How can we act when everything continually slips away, when everything is already out of date?

If politics is based on the 'distribution of the sensible,' what happens when the sensible is liquidated in favour of the reign of the present, which is nevertheless precisely the time of politics? If all politics is founded on the radical aesthetics that is the distribution of the sensible, then the advent of the hyperpresent has reconfigured the space of political possibility. The culture industry has constantly shown itself to be a strategic centre in this respect. In seeking to establish a monopoly over the given-to-be-heard and the given-to-be-seen, it establishes a hold on the sensible, itself obedient to the synchronous community, a community that denies the end of anything by sublimating the present into an ever-renewed eternal stasis. Through the hyperpresent and the reduction of the sensible to its synchronous function, the ability to see and to hear, or more exactly the ability to act on the basis of vision and audition, is gradually revoked—and along with it, the very possibility of a politics.

A paradox lurks in this epoch of intensified flows of information where there is always more to see and hear. The more rapidly everything moves, the more it freezes into indistinction. Such is the paradoxical reality of a society that prides itself on its commitment to living in the present, but which makes of the present a dead time, a state of perpetuity that regenerates itself through the capture and synchronous exchange of the sensible.

As we have seen, the 'killing' of time is not a derealisation. It is less a withdrawal of the real than a withdrawal of becoming (unless this is precisely what the real is—not a fixed point that is here and now, but the articulation of fixed points, their causal sequencing). And as we have said, death then retreats. It becomes an increasingly ungraspable horizon. The exercise of power no longer depends upon death, no longer delivers it as a sentence. Power no longer has any need for it, preferring to leave behind the menacing aura of death brandished as an instrument for maintaining order, in favour of a more direct hold over life procured through a monopoly on time. The terminal entropy that death engenders is itself beyond death. Death is no longer the ultimate limit. But the disorder it represents—the abolition of structures, the destruction and end of all things—endures nonetheless in promises it is increasingly unable to keep.

And it is against this terrible entropic promise hidden at the heart of finite-being that the synchronous community of projected-beings raises itself up, celebrating the reign of static, hypersequenced time, repeating and forgetting itself ceaselessly, simulating the conservative, protective eternity of an Optimum where nothing can ever happen.

Faced with the tautological coupling of a stamping out of the real and a ubiquitous amnesia, it becomes urgent to reactivate a thinking, not of objects, but of becomings. The toxic coupling that links finite-being and projected-being is precisely the hyperpresent. Its potency increases in parallel with the acceleration and intensification of exchange and synchronisation in the Human Network. The hyperpresent nullifies the present, for the present without shadows is an eternity. This is the perpetual present—a life sentence.

3

THE EXPANSE OF TIME

CONQUEST

But what about the time to come, the next stage? What comes *after* the present? We have lived through the first two decades of the twenty-first century and we still haven't managed to rid ourselves of inertia. As we are well aware, across the surface of the Earth there are fewer and fewer dead zones—places that remain unsynchronised with the Human Network. The Network is close to achieving its dream of fusing with the infinity of the world. But it is not everywhere yet. It still takes time to synchronise our 'connected' devices, and interaction isn't always immediate. Asynchronous spaces, or at least breathing spaces—intermediate zones that interrupt the hyperpresent—still exist: forgotten spaces, deserts, jungles, mountains, rural areas, chasms, or steppes. But these empty intervals are becoming rarer. They have already been compromised by the pansynchrony of satellites, and there can be little doubt that they will soon be wiped out completely, or at least rendered insignificant. Which brings us to the question: What comes after the acceleration of the present? Will ubiquitous stasis be able to sustain itself once it has reached the limits of acceleration? Or will it have to go on the offensive, seeking new spaces to conquer?

The present is already overextended. It has bitten off more than it can chew. The empire of the present wants to annex the past as well. It holds onto events so that they can no longer fade away. They are 'saved'—permanently imprisoned in digital memory. Data storage channels the flows of our lived experiences into a potentially eternal refresh loop. Even twentieth-century human beings are already partially undead. Photographs bear witness to their appearance, films and tapes have preserved their gestures and their voices. People die but they persist as illusions. A part of them survives, embedded in a pseudo-eternal time which is the time humanity wishes to make its own. Soon, this logic of preservation will also take hold of human memory, so that it might be restored at any moment. In this way the hyperpresent will destroy any possibility of dying, indefatigably perpetuating itself, tirelessly expanding, so that the real will eventually be absorbed into the actual, the current.

The great enterprise of synchronisation has subverted the disillusioned motto of the punk movement. *No Future*: a phrase that once signified defiance against order and social structures has now become a factor in their consolidation. The modern epoch was the epoch of projections, the post-modern era that of stasis and petrification. And the era to come announces itself as that of temporal cannibalism.

The present eats time as Saturn eats his children—out of fear that one of them will usurp his throne. Thus the instantaneisation of the present goes hand in hand with a will to total recall, where everything that happens has to leave a memory—that is to say, everything must be retained so that it can be infinitely resurrected. Once again, it is against the inexorable passage of time—and, through it, against our powerlessness to endure, to remain, to transcend our limits—that the logic of the hyperpresent militates. And it is the refusal of this constitutive impotence of our being that generates the sterile need to conserve and replay everything. To replay it not so as to live in the past, finding solace in the recollection of what is no longer present, but so as to absorb the past and digest it into the present.

All of this is a question of a force of attraction. The past is reactualised and the future no longer exists, or rather it no longer *speaks*. All that remains is the present, which acts like a black hole, sucking in projected-being—a social entity captured by the synchronous community, decoupling itself from the finite-being of which it is nonetheless an extension.

Decoupled being, in its contemporary form, is absorbed into the synchronous community constituted by the hyperpresent. The human of tomorrow, augmented, hypermnesic, accursed because incapable of not remembering, will ultimately be able to do away entirely with forgetting and the past, thanks to external memories interwoven ever more tightly with their perception of the real. Access to the past will become a procedure that no longer involves distance or duration, to the point where the past even loses its function of gradually retracting what was into nonexistence. The future human, a being whose date of arrival is still uncertain but who is already being called forth and promised by transhumanism, will transcend its biological death and bleed out into a multitude of eternal synchronous flows.

Eternity for all, and the definitive abolition of death—such is the destination of which humanity dreams. At first—from the very dawn of time—the negation of death was expressed through the concept of a beyond, found almost invariably across all cultures, and understood as an eternal world where souls are preserved, freed of the contingencies of the flesh. But now the denial of death is deployed along two parallel axes: in the eternal simulacrum generated by the hyperpresentation of the present, and in the struggle waged by technology against the physical and biological limits of finite-being.

Curiously enough, transhumanist prophecies, while serving the denial of death, belong to an already obsolete paradigm—that of a modernity bent upon the infinite extension of progress and scientific innovation. This modern paradigm of science as a discipline that endlessly revises itself, outdoes itself, and opens up fields that were unimaginable only a few decades earlier each time it surpasses its own limits, no longer describes the world of the hyperpresent. For humanity's horizon is no longer that once dreamt-of eternity of permanent progress. Humanity itself may not even survive the effects of science's century-long hegemony.

Whether or not it is recognised as part of this lineage, the transformation of the human is already underway. And it is risky, painful, and dangerous. Risky because the hyperpresent is precisely time without projection—a time that renders all conjecture impossible. So it is futile to ask whether it will stay *like this*, or how long this simulacrum will continue to function, how long humans will be able to bear the acceleration of everything in the service of instantaneity, and how long they can possibly endure the alienated suffering that comes with the decoupling of finite-being from projected-being. It is this decoupling that presents the real danger—the consequences of which are already beginning to be felt. The extraction of being out of itself, the replacement of the body, the contestation of the finite nature of the individual—none of this can go any further without a *simplification* of the human.

The ambitious enterprise of 'mapping' the human has already begun. But humanity is not a territory that is easy to model. Now, faced with a territory that is too complex, a cartographer has two options. The first is to revise the method, refine the analysis, and to begin the work again in such a way that it will do justice to its object. The second is to go to work on the object itself, reducing it, simplifying it so that it ends up becoming crude enough to correspond to its representation. It is this latter process that is currently underway, and in which humankind finds itself in the difficult situation of being both the cartographer and the terrain to mapped.

Everything is now presented to us in the form of simple choices, boxes to be ticked, whether it's a matter of demographic category, genetic heritage, or the political party that one feels the most affinity with—not forgetting, of course, the incessant demands of the Network, with its constant injunctions to tell it whether we like, or do not like, such and such a piece of information, such and such a story, such and such a flow.

Now, simplifying questions automatically simplifies the responses. Reducing the ways that sentiments and sensations can be expressed reduces the sensations and sentiments themselves. Thus, the reduction of the human being to a parameterisable model threatens to bring about a de facto reduction of human thoughts and affects that will keep existence captive, reducing it to nothing but what is communicable. If, as Nietzsche says, 'we cannot even reproduce our thoughts entirely in words',[3] then how can we hope to reproduce them in code?

To this authoritarian *expansion* of the present into the passage of time we might oppose an understanding, or rather an intuition, of time as an *expanse*—a changing landscape whose features begin to define themselves as one draws closer, or, on the contrary, start to lose definition as they recede into the distance, revealing different facets at every step of this retreat of the sensible—every step a resonance, a memory, slowly attenuated until it reaches the point of total forgetting that corresponds to the permanent disappearance of any trace.

---

3. F. Nietzsche, *The Gay Science*, tr. J. Nauckhoff (Cambridge: Cambridge University Press, 2001), 148 (§244) [translation modified].

Because people and things still fade and die. Erosion, the great sculptor of finitude, carries on its work. But we refuse our temporal boundedness just as we refuse our spatial boundedness. We want to be everywhere, all the time. We want nothing more than to locate ourselves in a nonplace and an untime. Society dreams of the permanence of things. It thinks it can secure its eternality through a fantasy of stasis. But its agents are finite, their existence bounded and situated. Civilisations crumble, transform, are destroyed, and eventually die. This is the reality, and we have forgotten it. The empire of the present has persuaded us, screening time out, impeding the operation of the hinges that join past, present, and future. Networked eternity negates what we fundamentally are: beings in becoming, beings whose existence is finite and will see us separated from one other in the end.

THE ENDS OF THE WORLD

The ancient concept of destiny deprived human beings of a certain freedom of action. Whatever they did, fate could catch up with them and derail their plans. This mechanism of impotence functioned principally as an organ of regulation and control. In the age of gods and oracles it was understood that there were designs at work that were greater than those of puny humans. Humans were mere playthings. This did however have the virtue of positioning each individual within a wider world—a world whose becomings exceeded them, but impacted them *locally* all the same.

This fatalism was opposed by the complete individualisation of human beings and their destinies through the notion of free will, which has rapidly become one of the principal weapons in today's war against death. Individual omnipotence is effectively expressed in freedom of choice and action.

And from this point on the sky is the limit. The warding off of death is achieved via the repulsion of our own limits and by the illusion of synchrony, which, through its permanent solicitation of our attention, induces the feeling that, so long as we are synchronised, we retain mastery over our destiny—that *anything* is possible and *nothing* bad can happen to us.

The hyperpresent has persuaded us of the *temporary* eternity of each and every one of us. We are all eternal, right up until the moment we die, because the actual process of dying is becoming less and less concrete. We pass directly from eternity into shadow with no transition, perhaps disappearing behind the thick curtains of some hospice or other on our way out. Death is no longer something that we embody. The rejection and suppression of finite-being—of mortality—is also manifested in the evolution of society's relation to the body, ever more inclined toward an administrative regime (from the management of illnesses, psychological problems and syndromes, to the regulation of hair growth, and even the body's transformation through cosmetic surgery, prostheses, and tattoos). This symbolic reconstruction of the body eliminates the very elements of precarity and ephemerality that are, precisely, constitutive of it. They are countered with hygiene, health, and youth—in other words, with an abstraction: eternity. The human body has become a glorified body extracted from all contingency.

The hyperpresent is narcotic, in that it induces a forgetting of self. It imposes an incomplete vision of the world in which the instant and the everyday play the role of a refuge, and where repetition and the marking of time function as rituals for exorcising the fear of death, a fear all the more terrifying now that it has been dissimulated.

The conjuring away of death has long been the task of two institutions, one sacred, granting access to the afterlife, the other profane, prolonging the existence of the dead through their lineage. Since time immemorial, and today still, a great deal of importance has been attached to dynastic logic, whether applied to an empire, an estate, or just a holiday house by the sea. Seeking comfort in the preservation of a legacy and the joy one imagines it bringing to one's descendants is a delusion, though. Such things will never amount to immortality. And even if we understand the latter symbolically through the nebulous idea of posterity, it doesn't change the fact that we must all, ineluctably, die.

Accumulating wealth in the hope of passing it on as a legacy seems just as naive and derisory as the more concrete attempts of those cryogenically frozen billionaires mentioned at the beginning of this book—patiently awaiting their reawakening in glass sarcophagi in the desperate hope of avoiding their fate. Our children can't help us. The fact that my genetic lineage can be transmitted doesn't mean that I will survive my own annihilation. For it itself is doomed to fragmentation and will inevitably come to an end, one way or another. There is no posterity that could possibly abolish my death. When I die, everything ends. Even if the memory of me and my ideas continues to resonate, even if I live on in the world a little longer through them, it still amounts to very little. For this will also dissipate sooner or later—those who have known and loved me will die in their turn, and things will gradually break down until the reddening sun finally ceases to warm a dying planet that humanity has still not managed to leave. As Proust wrote, 'eternal duration is promised no more to men's works than to men'.[4]

---

4. M. Proust, *Time Regained*, tr. C. K. Scott Moncrieff (London: Chatto and Windus, 1970), 524.

Even so, ubiquitous synchronisation promotes a powerful illusion of total solidarity between my finite-being and the world of the synchronous community. This illusory solidarity distorts my understanding of my own ability to participate in this world. It generates the belief or superstition that *my* end—that is, the end of my world, the end of my story—is The End of the World and The End of History. The assimilation of global, cosmic becoming to the reduced scale of our lives is a powerful sign of the domination of projected-being over finite-being.

How many false prophets have announced the imminent end of the world, invoked the end of History, or predicted imminent catastrophes and cataclysms? How many have tried to inscribe the end of all things into the contemporaneity of their own lifetimes? The excentring of self and the flight of worldly, projected-being into synchrony generate confusion within the individual, who now lives out a fantasy of themselves as a human-world. The world could very well end, a cataclysm could indeed befall it. But the motivation of those who invoke such disasters is to be sought more in the human-world complex than in objective analysis. The world is still here, although of course we can only presume that everything, sooner or later, will necessarily come to an end.

'Every true history is contemporary history', wrote Benedetto Croce.[5] History considers the past through the eye of the present, and thus risks cultivating the pretension that the whole earth can be encompassed in a single glance. But our 'temporal gaze' is exceeded on all sides by events that have contributed to the evolution of clades, built civilisations, and brewed climates.

---

5. B. Croce, *Theory of History and Historiography*, tr. D. Ainslie (London: Harrup, 1921), 12.

Believing in the end of history or the latest end of the world is no less illusory than believing that we can see grass growing with the naked eye. The temporal regimes within which these events unfold are not the same as those that govern our experiences of living. To say that your epoch is a decisive one is just a naive way of attempting to make yourself intemporal—that is to say, immortal. Historical fact has only a local function. Its resonance is limited, finite. To anticipate the end of everything is also to no longer feel alone in the face of one's own death.

But the fear of dying isn't a private sentiment, it's a global fantasy—a ubiquitous neurosis. And it is of course based on a concrete fact: the inevitable death of each and every one of us. The synchronous society has tried to hide death, to push it back into the dreadful shadow of the void. We have never really known how to face the end of ourselves and of everything. We have never had the courage to completely deny paradise, that providential place where everything is resolved in the best possible way, and yet we have never been able to fully subscribe to the idea either, leaving us with a niggling feeling that something has been left unaccounted for. So the representation of an ultimate end of all things has been chased back into the limbo of existence. And an impossible space has been prepared for it—the space of terror.

A community that dreams itself as eternal, instantaneous, and infinite is a delusional community. This is the exact purpose of the synchronous Network—it is the synchronous community's communalisation of the present that is the source of the forgetting of personal death. This kind of community, caught up as it is in denial, is incapable of forming a politics adequate to what is effectively the fundamental requirement of humankind. Not a requirement for consolation in the face of one's inevitable death, but on the contrary, the need to be confirmed as mortal.

Faced with the abstract and eternal world of projected-being—a world of signs—it becomes necessary to reaffirm the sensible, bounded, limited world in which finite-being develops and in which it cannot be dispersed. For if finite-being can itself be considered a multitude, if it is in a certain sense an infinite world made up of an incalculable number of isolated elements placed into relation, it nevertheless resides in that strange unity that is the individual. I am the one 'who' is finite. I am the place-holder of a finitude.

A knife stuck in my flesh doesn't simply result in a transfer of energy accompanied by a molecular rearrangement of metallic compounds and organic tissues. It makes *me* suffer and it spills *my* blood. Finite-being is a capture, a captivation of a matter drawn off from the infinite world and determined by a common becoming. And yet an individual cannot be reduced to its radical being.

There is an articulation that is operative at every moment of our existence, a hinge between our own becoming, which is necessarily precarious, and the social, cultural being that animates us, that defines the greater part of our actions, that governs the greater part of our affects and which finally takes over, sometimes to the extent of being mistaken for our own existence.

From this point on, projected-being, swept up by synchronous forces, begins to develop imperialist fantasies. It wants to be eternal, omnipresent, and omniscient. It sees itself as having transcended finite-being. But this is not the case. Pain, pleasure, and joy are still experienced through a body that expresses them in a way that no Network cartographer yet knows how to model.

THE SHADOW OF THE PRESENT

The regime of stasis nurtures a mode of being that is mesmerised by objects and their appearance. It succumbs to a runaway process, trying at all costs to keep apace of events so as not to be distanced from the world, becoming trapped, like all the rest, by the hyperpower of a currency that sells itself as the sole source of impact upon the real. But the dominion of Stasis, the instantaneity of the present, is not an inescapable fate. It is a will—a diffuse and multiform will, but a will all the same. And as such, it can be countered. Other understandings of the present are possible, ready to contest its empire, ready to put a stop to the implacable mechanics of the current. There is a whole network of actions that can be unleashed upon both flanks of doubled being, instigating a strategy of jamming, the first spark of a struggle against the folding of everything into everything, against the triumph of the current and the great tautological indistinction.

Modernity made the present into the moment where future promises are seeded through the doctrine of progress. Postmodernity proposed a reading of the instant as proliferative, a sheaf of multi-signifying becomings that undo the 'arrow of time'. But neither of these paradigms orient our experience of the world here and now. There is no longer any real projection, no longer any putting into perspective. Everything is decided and unfolded through a hyperpresent that reveals nothing beyond itself. The hyperpresent is the epicentre of a great pyre of consumption and consummation. But the present is not the hyperpresent. It cannot be reduced to the instant. It has a duration of its own which is that of its resonance. For the present is always accompanied by two shadows, cast by the projections of two suns—one future, the other past—between which it stands and to which it constantly refers.

We do not live in the present, still less in the instant, but much rather *in the shadow of the present*. The present is the knot through which past and future resonances can express themselves and expand. It is the temporal space in which the future is glimpsed and the past can flourish and sow the seeds of the contemporary world—the complete inverse of the logic of updating the past by refusing it its original nonexistence (must we remind ourselves that the past *is no longer*?). Such a logic is a logic of death in so far as it denies life by refusing its precarious, uncertain, and terminal character. It denies life because the living are destined to die, and denying the mortal component of living things amounts to draining them of their intensity, replacing life with a smooth, cold, mirrorlike surface.

So there is a labour of transvaluation to be carried out here. We must rediscover the thickness of the real and no longer just content ourselves with its surface, which is only a surface of inscription, a set of communicable and synchronisable data. This surface must be scoured in order to excavate the real that lies beneath. It is then a matter of establishing a relation to the world that resists the pressure of hyperpresence, a relation that is turned toward the current and the outside, which are both resonances of it. We must reinstate the presence of death in the sequence of life and accept the fact that we are destined to die.

To live in recognition of your own mortality is precisely not to glorify the present instant, as the contemporary hedonistic interpretation would have it. It is, rather, to constantly be mindful of the expanse of your life and to place it into perspective against the epochs and centuries that have gone before it. Not so as to render it insignificant by immersing it in a flow of becomings that will build into a rogue wave, poised to inundate everything in a maelstrom of obscurity, but to rescale everything—the chain of events, universal history, geology, even climate change—to the time of limited lives, to a succession of lives, boundedness within boundedness, some already gone, others yet to come. It is to stand up against Eternity, against the permanence of things. To reignite a kind of thinking attuned not only to objects and states, but to flows, processes, and strategies. A thinking of the distant horizon, of things that pass away and come into being. Objects must be reinscribed with their becomings, and we must recognise that objects themselves are in fact no more static or permanent than solar systems or nations. That there are only locally and conventionally fixed representations of processes—or things in becoming. And therefore we must accept the impermanence of the world and the beings that populate it.

This type of thought, borne by becomings, projects itself everywhere and into all times. It is not restricted to the past and to that which is no longer. It permits anticipation and speculation. It questions history and calls ancient times into question, for it has the upper hand over them. It releases dual being from the grip of the present, compelling it to accept the temporality of life and its finitude, and restores the fecundity of the articulation between finite-being and projected-being, thus opposing itself to the nihilism and incoherence of the present. In this way, it disarms a certain thinking of the instant in favour of a thinking of a world in becoming, forever reaching toward the before, the after, and the elsewhere.

We have been prohibited from living with nostalgia, melancholy, and the fear of death. They have been banished from the empire of the present. But they are important, because they contribute to the dual individual's understanding of the constitutive difficulties of their own doubling. These sentiments linked to becoming are the transversal functions that alone are able to articulate the asynchronous temporalities of the political and the existential.

Melancholy, for instance, is the quicksand you sink into when you can no longer believe in reality, and no longer know what to look for in fiction nor what to make it say. But it is also the sign of an internal confrontation between the actual and the potential, between what is and what is yet to come.

Sehnsucht, *saudade*, *wanderlust*, *fernweh*. So many foreign words designating an understanding of becoming: nostalgia, melancholy, desire to see the world, longing. We need to rediscover a certain idea of romanticism and situate it in relation to our networked postindustrial world. We need to cultivate enough courage to no longer deny the sadness and exaltation of being in the world, of feeling minuscule and impermanent in the face of immensity and eternity. With our eyes wide open, we need to confront our finite being with the enormity of the world's, and call forth a thinking of distant horizons and elsewheres.

Through melancholy, *sehnsucht*, *fernweh*, and *wanderlust* is formed a relation to the world directed toward absences, projections, and phantasms just as much as toward objects and beings. The desire to see the world is a feeling before it becomes an exhortation. It is a call from elsewhere, a force that draws us into the unknown. As such, it is a kind of antitourism. Tourism is the repetition of the same ritual, always similar, just *differently* repeated. Which doesn't mean that it is a ritual of difference. Quite the contrary—it is a ritual of identity that depends on a difference in implementation so that, by contrast, it can accumulate and be capitalised upon. In other words, it serves the wordly being's dream of omnipresence, of being everywhere all of the time.

The experience of the immediate, of the here and now, cannot function without the affects that accompany becoming—those elsewheres and beyonds that are forever moving through us. They are traces of the real world that we have lost, eddies that stir up the ruins upon which the present is built. And these traces, these uncertain affects, illuminate the depths concealed beneath the surface of the present. This may be the hidden ulterior principle of places, streets, and laws.

We are continually shot through by a multitude of times and places that coexist alongside one another and are woven into our experience of the present. To say this is not to deny the real, or to suggest that it is nothing but an empty shell with no actual substance, particular to the moment of its appearance. Instead, it is to reestablish the *irreal* (the past event, the event to come, the event that could have taken place, or which has, perhaps, taken place elsewhere) as a *modulation* of the real that would be mute if it were nothing but itself.

This modulation should be understood as a decentring function: it helps us to resist the pull of the present. It reminds us of our own death, of things that have been lost or forgotten, of disappointed loves, and of great loves that are no more. It is a plunge into a deep well that gives rise to a process of differentiation. Nostalgia has this function: it is an experience, sometimes a painful one, of becoming. It is the experience of something that no longer is and that one wants to bring back. But it is also an experience of expanse, of a vastness in which meaning and direction coincide, and this is why it is so sweet.

It leads us elsewhere. And elsewhere hardly exists anymore. Synchronisation has gradually destroyed it. Everything conforms, everything harmonises, because everything is synchronised.

The synchronous community has dictated its own rhythm to us, the vertiginous stupefying rhythm of the saturation of present instants. But the placing in common of the present via globalised systems of exchange and information and their omnipresent personal and mobile receivers does not for all that clearly determine a collectivity. Such a collectivity could not exist. '*The* community doesn't exist. There is only community.'[6] And here the imperative of permanent universal synchronisation collapses under its own weight. The placing in common, or sharing, of a space of communication and values is then no longer an imperial gesture, but once more becomes a *means* of common experience. And so it can be replaced by an asynchronous common space.

The asynchronous space of community rebalances the disequilibrium between projected-being and finite-being. It is a space in which the privileging of one at the cost of the other ceases, or at least is attenuated. Concretely speaking, this space has always been there, even if it is now drowned beneath the synchronous flows. It is the shared space of birth and death, a space defined by the placing in common of the reality of our limits. It is the inertia of time zones and circadian rhythms. Being asynchronous means living in the inertia of the real, inhabiting the latency between moments of actualisation. But this latency isn't futile or sterile. It is fertile, because it is latency that makes room for the intermediary time in which becomings take shape.

---

6. Tiqqun, *Introduction to Civil War* (Los Angeles: Semiotext(e), 2010).

Asynchronous space is a multiplicitous space—an ensemble of space-time localities, isolated by definition, patchy, secret, and hidden from the law of instantaneity. These are remote geographical zones, as we have said, but also human spaces, laboratories and enclaves whose activities cannot be conditioned by the imperatives of the hyperpresent. Desynchronised zones that have their *own duration*. Heterotopias—or better, *heterochronias*.

Time is a dynamic dimension that can redistribute the given, fold the known back into the unknown, and project uncertainty into the assured. Time thus becomes the vector along which to recast the dice that configure and determine the real. A complex temporal relation is then fabricated in which, for example, the present can be understood as the past of the future. '[O]utsize buildings cast the shadow of their own destruction before them', writes Sebald in *Austerlitz*, taking the example of the Palais de Justice in Brussels; '[they] are designed from the first with an eye to their later existence as ruins'.[7] A methodical and patient archivist, Sebald observes the tremors of past and future worlds from the standpoint of the present. In his work, the strange, the forgotten, and the residual cut through the very banality that produces them. The apprehension of the actual, of the everyday, is never allowed to mask this conviction that the surrounding world is condemned to one day disappear, that everything is precarious, but that we are standing, here, now, atop nonexistence, as if we had been superimposed like a transparency upon long lost ancient worlds. Thus, nothing can really be foreseen and nothing is given forever.

---

7. W.G. Sebald, *Austerlitz*, tr. A. Bell (London: Hamish Hamilton, 2001), 23–4.

This conviction is exemplary of a stance both restless and serene, in which speculation is capable of redefining our relation to the world and the present instant, and of redeploying our existence no longer within a Network of confirmation designed to console projected-being, but in a barren place, battered by the winds of vastness and nothingness, encircled by the void.

If death is one day vanquished, we will still need to think according to our own limits. If the day comes when we no longer have limits, then we will be able legitimately to lay claim to a society like the one we have today, for our reality will finally be compatible with it. But not beforehand. What will then emerge will be a world designed for the new gods that we will have become—an immense Olympus, where all that remains to assuage the boredom of Eternity will be secret plots, the delirium of power, and the annihilation of our enemies.

Until that day comes, we must live in the shadow of the present, which is the shadow of death.